Goodbye, Darwin
by
Art Cooper

Key Publishing Company

3240 Pine Grove Avenue
Port Huron, MI 48060
Phone: 810-984-5579

Typeset by Walkabout Ink
Port Huron, Mi 48060

First printing—May 1994
Second printing—November 1994
Third printing—February 1995
Fourth printing—November 1995

ISBN # 0-9650491-0-8

Acknowledgement

This is my opportunity to thank the Reverend Bill Hossler, without whose support this book would never have been completed.

Bill's constant enthusiasm and encouragement provided the essential energy and drive so that an old chemist was finally able to finish the work God had set before him.

Art Cooper

Contents

CHARTS

Author's Forward

This is not a science textbook. Make no mistake about that. Even though trained in math, chemistry and biology, I have made no attempt to offer a step-by-step rebuttal against the theory of evolution. That would be possible, of course, but also long and boring.

Instead, I have chosen the key tenets of Darwin's thinking, his basic assumptions, and have attacked these, using newly-discovered evidence. This is recent data which was not available in 1859. If Darwin had been aware of these things, he might have dropped the idea of organic evolution altogether, or at least a large part of it.

Frankly, much new and illuminating information has come to light in recent years, information which strongly suggests that Charles Darwin's theory was in error. Unfortunately, these findings have not reached the young adult reader via scientific magazines or publications, primarily because many scientists are anti-creationists and/or atheists.

Therefore, the purpose of this work is to provide young adults with a quick reference, a hand book of facts which challenge Darwin and his theory of evolution. I have made no attempt to embellish any of the enclosed data, nor have I bent any sections of the Holy Bible to fit my own thesis. I have simply presented the information as a set of facts which supports that our planet and ecology did not evolve. Clearly, if the earth did not evolve, we are left with the alternate option: Our world was created by God.

I hope you will read each chapter straight through. Pause occasionally to think about what you have read. Keep your Bible nearby for the last three chapters. If I can bend your thinking or call Darwin's theories into doubt, the writing of this little book will have been a total success.

Chapter One

NOW, HERE'S DARWIN

y first acquaintance with Charles Darwin occurred in the autumn of 1954. I was in the 8th grade, and every student in our school took a class in Basics of Science, which was taught by Mr. Norris. As part of the course, we were introduced to the theory of evolution and its originator, Mr. Charles Darwin. Having been raised by church-going parents, I was shocked by Darwin's views, since they were essentially godless. It was clear to me, even at age thirteen, that to accept Darwin's teachings implied that man is nothing more than a very intelligent animal. My mind, although young and formative, recoiled from such a view, even though the evidence seemed stacked against me.

As I progressed through high school and college, I found that both Darwin and his theory of evolution were widely accepted. In fact, this acceptance was so universal at the college level that no one seemed to question its validity. Yet, my mind still struggled. I felt as though I was the only one who noticed a few flaws in the proof of evolution. My professors smugly assured me that Darwinism was a fact of life, and frequently made fun of my questions. However, as time passed, I became more convinced that I was right. I saw serious errors in Darwin's original assumptions. A great deal of data was missing or nonexistent. Other areas of science seemed to contradict the proposed mechanisms of evolution. Even so, I recognized that Darwinism had become embedded in our colleges and universities as part of an overall philosophy, the same as Marxism and many other popular 'isms.' Most science professors, in order to protect their jobs, did not want to rock the boat. Darwin became part of the System, unchallenged by the intellectuals.

I fully realize that many young adults today are facing the same situation as I did during high school and college. It is risky to question or contradict one's instructors. After all, they hold the power of grades. Further, it is the teacher who usually chooses textbooks and course materials which reflect his or her own views. Some universities actually prohibit the presentation of Creationism in their science curriculum. God has been essentially banned from the college campus in many northeastern states. Reliable books which directly compare evolution and creation are virtually impossible to find in bookstores.

So let me pause here to draw the line clearly for the

reader: the theory of evolution is NOT a proven law. It does not have sufficient data to support it. In fact, many high ranking Darwinists in the science field are becoming concerned about the holes in the theory. Their concern is well founded. Things have changed radically since my days at Garfield Junior High.

Fortunately, I am an industrial chemist, fully degreed and yet totally independent of the academic culture. As such, I can speak out fearlessly. No one can take away my job or my seniority because of what I say here. So read on.

The year is now 1995. Forty years have passed since my first science class with Mr. Norris. Many new discoveries have emerged, particularly in the areas of physics and chemistry. Computers have also greatly aided the advances in genetics, and the stunning discovery of DNA has revealed the workings of genetics. Geology has moved forward in its studies of the earth. Men have visited the Moon. Space is no longer an unknown as we routinely see news clips of astronauts crawling about on drifting shuttles. Times have changed and so has science.

In this short book, I intend to show why the theory of evolution is no longer valid. It can no longer be considered science, but rather a form of godless religion. Those who teach evolution are, in fact, turning their backs on recent discoveries, or worse, are ignorant of them. Darwinism could be classed as a cult, a curious club of atheists with peculiar views, rather than a part of the mainstream of science.

There is but one source of man's existence: God.

Read on.

Chapter Two

A NEW THEORY

he Theory of Evolution was held to be the scientific answer to the creation myth. Charles Darwin, who had been a student of theology himself, saw evolution as a replacement for the Bible creation account in Genesis. He saw this replacement as a struggle between pure science and old religious doctrine. There was, in fact, a struggle since so many godly people denounced Darwin's theory, calling it perverse, twisted, and devilish. However, the theory of evolution, according to Darwin, was not based upon lies, but facts. He wove an interesting theory to account for the origins of all things. His book, *The Origin of Species*, was published in 1859 and revised in 1880. In this work, he put forth the main assumptions of his new theory:

1. That the Universe itself is timeless and endless, having existed forever with no beginning nor end.

2. That, given such enduring conditions, one or more planets had eventually developed ideal environments.

13

3. That, within an ideal environment and endless time available, life had gradually arisen without direction, i.e., according to random chance.

This is the Theory of Evolution in a nutshell, a scenario which allowed for unlimited chance happenings and failures. Under such a set of conditions, the strongest or most fit individual would ultimately emerge. So Darwin boiled down his thoughts into one key phrase: 'Survival of the Fittest.'

Theory vs. Law

Around the time of Darwin's book, scientists of the day began to resolve some of their own questions by organizing scientific information into two categories:

1. **Laws**: These were the principles which were proven beyond a shadow of doubt. For instance, there is the Law of Gravity. The earth exerts a pull on every object on its surface. If you were to jump off a ladder, you would be pulled directly back to the ground. We call this falling. Actually, it demonstrates a scientific Law which works every time. There are many such Laws in science.

2. **Theories**: These were those principles or suggestions which could not be proven by existing facts. A theory is a possible explanation for a set of circumstances, which later may be proven correct.

For example, Albert Einstein (whom we will discuss later) introduced his Theory of Relativity in a day when it could not be easily proven. Since then, most of his theory has been proven and has become the Law of Relativity. All theories must be proven by concrete tests, which require enough facts (or data) to substantiate them. Otherwise, they are discarded.

The scientists of the 1880's were careful men. Since Darwin's ideas were not well supported by facts, they recognized that evolution was simply a theory. Essential key pieces of evidence remained to be found; thus, they could not classify evolution as a Law.

Darwin was totally confident within himself, and, in later writings, claimed that his theory would be fully substantiated by discoveries over the next 100 years. If not, he said, he would have to seriously question his own theory.

Darwin and Lyell Charles Darwin was a naturalist, an observer of nature, a student who, today, would be called a biologist. He knew almost nothing about chemistry, mathematics, or physics. He had only a very basic grasp of geology, or earth-science. His grasp of astronomy was rudimentary at best. In fact, his skill in studying nature was largely self-taught, since his areas of study at Oxford University in England consisted of Christian theology and philosophy. However, Darwin was a clever writer and a good speaker. He also was a wealthy aristocrat, which gave him significant connections to people in the British Government. These connections helped him become appointed as Royal Naturalist aboard the naval vessel, Beagle, whose voyage around the world was to be one of scientific exploration.

Upon returning from his journey on the Beagle, young Darwin began to formulate his ideas about evolution, based upon observations made on the trip. However, concrete evidence was quite lacking. He had found no solid data which could show a gradual change or improvement in various creatures. Neither did he recover any living specimens, whole skeletons or fossil chains. His new theory hung upon his own subjective observations.

At this point, he was aided by his friend from Oxford, Charles Lyell, who was an amateur geologist. Young Darwin began filling in the gaps in his theory with information gleaned from Lyell. Soon, his completed book, *The Origin of Species*, was published. His newly printed theory was full of assumptions and glib assertions. It was easy reading, since it contained very few charts or tables of data.

Reading Charles Darwin's original book was a great surprise to me. It is nothing at all like a science book; it reads like a novel. *The Origin of Species* makes a great tale, well told in a rather smug, self-satisfied style. But scientific, it is not! I was about 25 years old when I first read it, and, from that point on, I began to suspect that Mr. Charles Darwin was a master salesman who had sold his theory using a clever style rather than hard facts.

What were the basic scientific discoveries of Darwin's book? In my view, he claimed three things:

1. Given endless time and ideal conditions, life had arisen spontaneously from a simple one-celled organism. Over the countless ages, this organism had gradually changed (or evolved) into more complex multi-celled forms of higher life.

2. The most fit of the organisms survived and grew stronger, a process Darwin called natural selection. Further, these organisms altered their bodies to deal with their surroundings by a process called mutation, which spontaneously gave rise to higher forms of each organism.

3. Proof for these concepts would be found captured within the earth in the form of fossils, mineralized remains of these various creatures. In time, millions of these intermediate organisms would appear, producing a record of the step-wise alterations of each species. These fossils, then, would furnish the missing links in the chain of evolution.

In 1859, the above scientific assertions were startling and daring. They captured the interest of many intellectuals in Europe, becoming topics for many coffee-house

17

discussions. After Darwin's revision of his book in 1880, his theories were taught as philosophy, but not as science.

Karl Marx: By 1900, one European had taken note of Darwin's work, and made evolution a key point in his own ideas on Communism. Karl Marx reasoned that man was a rational, thinking animal who needed only the ideal social structure to advance. Darwin's theory provided a significant basis for Marx's own writings. Thus, Marxist doctrine fully embraced the main concept of Darwinism, namely, Survival of the Fittest. In time, all Communist (and Socialist) countries accepted the Theory of Evolution as a basic fact, even though it remained unproven. Darwin had removed God from nature, and Communism endorsed this creed as part of its own belief.

Today, the philosophy of Communism is dying. Can Darwinism be far behind?

Read on.

Chapter Three

TIME, LIGHT AND ENERGY

he Twentieth Century brought very rapid gains in science and technology: automobiles, airplanes, rockets, computers, radio and TV, medicines. The list goes on and on. The advances have been remarkable, but, probably the greatest advances have been made in the area of nuclear physics, that is, the study of atomic particles and atomic energy. For here, in the realm of the atom, scientists began to recognize the unbelievable power harnessed within matter existing around us. As physicists began to unravel the puzzle of nuclear energy, a new question emerged: Where does all that energy come from? What force holds it locked in place, like an obedient puppy?

The answer was quick in coming.

Albert Einstein: Once in a hundred years or so, a special genius appears on the scene. Einstein was such a man. His mind was like no other, before or since. He had an amazing ability to understand the workings of the Universe. He grasped

19

the complex mechanical motion of stars and planets, suspecting that the Universe was not timeless or limitless. He realized that all matter had an age and that it was essentially made up of tiny energy particles, atoms, held together by a tremendous force. These atoms, in turn, were the building blocks of everything we call real or solid. Then Einstein did one more thing: he linked it all together into a mathematical equation:

$$E = MC^2$$

- OR -

Energy = Mass x (Speed of Light)^2

Few people, even today, fully comprehend this formula. Einstein described that Mass (or matter) consists of a huge amount of energy compressed into a small size. Conversely, this small amount of matter yields a huge amount of energy.

In one stroke, Einstein spelled an end to Darwin's first assumption that the Universe is timeless and ageless. His equation made possible the actual calculation of both measurements:

1. Time: According to Einstein, time is not linked to the rotation of the earth, i.e., 24 hours = One day. Time

can be measured by atomic speeds, linking it instead with thevelocity of light (C = Speed of Light = 186,000 miles per second).

2. Age: Since all Mass (or matter) is convertible into energy, it is very likely that it had existed (at some time past) entirely as pure energy. This means that, far from timeless, the whole Universe has a time span. Thus, the age of the Universe can be measured.

Most people have difficulty with Einstein's theory, since they tend to think of time as hours on the dial of a clock. We all know that a day requires one full turn of the earth on its axis; that is, one day and night taken together. One year, then, equals 365 of these days. Our sense of time centers on these familiar concepts. We normally calculate time using terms such as days or years.

Somehow, Einstein broke away from that restraint. He saw time the way God must see it: as a measure of mighty nuclear events. He realized that this Universe is an immense collection of energy, held in a tight pattern, and traveling through a great vacuum at a high rate of speed. Such a magnificent thing could not occur by chance, he said; it is the work of some Grand Creator. Having said this, Einstein died.

The Big Bang: Edwin Hubble was an astronomer. He was fully aware of Einstein's work, but was not attracted to nuclear studies; he preferred to be a star-gazer. One day in 1924, he noticed that photos of distant stars showed a change in the red color-spectrum. This was called the red shift, indicating that the objects were moving rapidly away from the observer. Hubble became excited. All the distant stars showed this red shift; they were moving away from the earth at different rates. After careful calculations, he deduced what this meant: The entire Universe was expanding outward, like shrapnel from an exploding hand grenade. Further, moving back through time, Hubble could see that all the stars had emerged from a single central point. This idea he called the Big Bang.

Since 1929, continuing experiments have proven that Hubble was correct. The entire Universe originated at a single spot in time. In a single instant, it developed from one stupendous explosion. The latest piece of evidence, documented in 1991, consisted of the discovery of background radiation from the Bang, spreading out much like ripples caused by a rock tossed into a pond.

Hubble estimated the age of the Universe to be 15 billion years. Recently, by careful computer modeling, this figure has been cut down to 11 billion years. Regardless of whether this is really accurate, the important thing is that the Universe is not timeless; it does have an established age.

The earth itself was, of course, a planet which appeared much later than the Big Bang. Our planet is not a star; it is rather cold by comparison, making it much more hospitable.

The surface of the earth is cool and covered with water. It also has a blanket of oxygen and nitrogen surrounding it. Modern calculation estimates the earth to be about 4 billion years old, but not habitable for about half that time. Therefore, we will say that today's best guess for earth's age is 2.5 billion years; that is, the time during which life might develop.

Please note that I do not personally endorse these ages as absolutely correct; I believe that the actual times are probably much shorter. Using the findings of Hubble (which are now considered a virtual law by the massive supporting data), I want to explore whether evolution would really be possible within that time span.

In any case, Einstein and Hubble have changed forever the previous concepts of time, light, and energy. They have defined the form of the Universe, as well as its manner of formation. They have provided a solid estimate of its age.

Could the Universe have occurred by itself ? What was this sudden source of immense energy for the Big Bang? How were all the stars and planets organized so perfectly that we can set our watches and calendars by their motion?

Read on.

Chapter Four

MODERN CONFLICTS

hat does the Bible claim regarding the origins of the earth and the Universe? Actually, the Genesis account describes creation using the same sequence as the Big Bang:

Genesis 1:1-3 KJV

1. In the beginning God created the heaven and the earth.

2. And the earth was without form, and void and darkness was upon the face of the deep. And the Spirit of God moved upon the face of the waters.

3. And God said, Let there be light: and there was light.

Here then are the events in order:

1. At first, there was nothing we could describe as solid, only an empty void.

2. At some point, God focused His incredible power, and the Creation of the Universe began abruptly, i.e., the Big Bang.

3. Early on, the cosmos was formless, a murky soup of raging fluids and gases, boiling in total darkness.

4. God's own intelligent direction took over at this point, causing the rampant energy to crunch down into little bundles called atoms.

5. Immediately, light became a separate quantity, flashing across the open spaces between bundles of matter "and there was light."

The only difference between Hubble's Big Bang Theory and Genesis One is a single item: Genesis claims that the source of power and direction came from God, while Hubble does not. Otherwise, the two accounts are very close and correlate well.

Of course, there is a second basic difference: the Bible claims that everything was made in six days, with a seventh day reserved for rest by God. Hubble's work suggests that creation took roughly 11 billion years. This difference appears to be too great to be reconciled, even though many written works have attempted to do so.

This seeming conflict was disturbing to me for many years. I was satisfied that creation was given a time frame, but the Bible seemed so far off on its description. Most Christian writers insisted that the days listed in Genesis were 24 hours in length, making creation take place within a span of 144 hours. This did not seem logical. Then I read a book by Dr. Ronald Youngblood, who pointed out that the first three days of creation were not solar days, that is, caused by the earth's rotation. The sun and moon had not even been created yet. This is according to Scripture, not Youngblood. The solar system was formed on the fourth day, during which the sun and moon took their current positions. Therefore the first three days were not necessarily 24 hours long, and the fourth day may or may not have been the standard 24-hour day. In other words, from God's own description, the first three or four periods of creation were an undetermined time span. When I saw this, I was greatly relieved. Quite clearly, it was possible that Christian writers had failed to interpret the factual account in Genesis, relying instead on a preconceived set of notions.

Still, the Bible version seemed to be unexplainably short.

Time Dilation: Human beings tend to be creatures of habit. We think of time as based on one full rotation of the earth, broken into 24 segments which we call hours. The full rotation we call a day. One complete orbit of the earth around the sun requires 365 days (or 8766 hours), which we call a year. All of these measurements are fixed, unchanged since the time of the Romans. All the nations on earth use this system. No one calculates time any other way.

However, time is not really fixed. For shorter periods, such as 100 years or so, time is essentially constant, so man cannot see significant changes during one life span. The first human to realize that time is flexible was Albert Einstein in 1905. Einstein stated that, when one approaches the speed of light, time will change. It may accelerate significantly as the speed of light is reached. Another way to say this is that events near the speed of light actually happen much faster than they appear to happen.

This effect is called time dilation. It is the compression of time, in which everything happens faster than normal. This effect was discussed fully by Dr. Gerald Schroeder in his recent book, Genesis and the Big Bang. Dr. Schroeder indicates that, as the universe expanded after creation, many things were happening all at once. No single measurement of time could be applied using our common units of days and years.

In other words, those early events of creation were controlled by God. They happened very quickly, with velocities very close to the speed of light. Measurement could only be made in fractions of a second in some cases. There is no way we could say the cosmos was formed within a specific time frame, i.e., 24 hours. Time was greatly accelerated under those circumstances, clearly a case of time dilation.

Only God knows the length of time it took to create our Universe. By current calculations, it appears to have taken approximately 11 billion years, but it may have taken less considering that time was dilated during those events. God told Moses about creation, probably while communing with him on Mount Sinai. Moses, a man who lived by the Hebrew calendar, wrote it down using terms he understood.

Read the account in Genesis One again; this time notice that the first three days have neither sun nor moon.

Then read on.

Chapter Five

THE FACTS OF GENETICS

So what have we learned thus far? First, we have learned that the conditions included in Darwin's theory, namely, an ageless Universe with unlimited time for the development of life, do not exist. The Universe as perceived by Darwin was a figment of his own imagination.

Secondly, we have learned that the Universe began at a single moment in time, in which energy was compressed into matter, and light sped across vast distances for the first time. This is not merely the Biblical account, it is also a widely accepted scientific theory, with sufficient proof to be accepted as law.

Really, the two points above are enough to bring Darwinism into question, since the entire foundation of evolution consists of a long and uninterrupted series of random events. Already, we have seen how unlikely this is, based on recent discoveries which Darwin did not have in his lifetime. However, there is far more evidence we must examine, so let's

29

go on.

We know that after the earth was formed, at some point, it became habitable. Its position in the cosmos was just right to support life, not too far from the sun, nor too close. Temperatures were very favorable, ranging from well below freezing at the poles to about 100°F at the equator. The surface of the earth consisted of 80% water, giving rise to clouds and rainfall (or snow at the poles). There was an atmosphere of air, made up of oxygen, nitrogen, and carbon dioxide. All of these factors came to exist in more or less perfect balance, providing a hospitable climate for life.

Recent studies tell us that the earth itself is a maximum of four billion years old, as we noted in the last chapter. It has been habitable for about half that time. That is, some forms of life could have survived, although full ecology, as we know it, would have taken additional time to develop. If we look again at Darwin's theory, we see that in the evolutionary scheme, simple organisms, such as the single-celled amoeba developed first. These, then, would mutate to become more complex over the passage of time, gradually evolving into higher forms, that is, the animals of our day.

There are two serious flaws with Darwin's scheme:

1. In order for this evolution to occur, there would have to have been millions of tiny modifications taking place, each one advancing the organism slightly. This

gradual process could not have moved rapidly, since each mutation was random and unguided. However, in the last 50 years, science has calculated a time limit of 2.5 billion years on organic evolution. It is doubtful that all of the necessary mutations could have taken place within that span. We will examine this idea later with an example.

2. If evolution of the species had really taken place, there would have been millions of variations of the same creatures in this chain of events. As the earth slowly changed, many of these creatures would have died out as unfit. Their bodies would then stack up like cord-wood in ponds and pits, leaving bones to become fossils. They would be found everywhere. The soil should be thick with them. After all, it is possible to still find the bones of dead cattle left by the pioneers. Surely then, countless numbers of previous life forms would be obvious to us as the Fossil Record which Darwin anticipated. Has it been found? The answer is, 'no'! Only an intermittent fossil record can be found. We will discuss this later in more detail, since it is very important.

Life: More than 100 years have passed since Darwin brought out his revised book in 1880. In his day, nuclear physics was virtually unknown, and paleontology (the study of fossils) was in its infancy. Darwin declared that life somehow arose by itself, without assistance, within the warmth of the sea, then proceeded to develop. Yet modern technology has utterly failed to generate life in hundreds of controlled experiments. In fact, scientists cannot yet explain why living things are alive. They can describe the traits of a living thing, but they cannot duplicate any of them. Most scientists realize that life is a wonderful force which is passed on from parents to the offspring. Where did this life force come from? Darwin, lacking an explanation, passed over the question rather quickly in his writings. Today, we are still dealing with it. Life began at some instant, just as the Universe itself did. When was that instant and what caused it to happen?

In our current day, naturalists are known as biologists who are divided into two groups: Those devoted to the study of zoology (living animals) and those devoted to the study of botany (plants). Although plants and animals are quite different, both seem to be alive. Two types of living organisms, then, inhabit the earth. Darwin concentrated his efforts on animals, looking for similarities among the species as proof of evolution. He ignored the whole area of plant life. This was an oversight, since plants also would surely have mutated in much the same way as animals.

Biologists find the plant kingdom most interesting. Surprisingly, plants have proven very resistant to change or modification. A corn plant always produces an ear of corn; it cannot be forced to produce peas or tomatoes. Attempts to make such changes always results in the death of the plant. There are no exceptions to this. In other words, plants cannot be made to mutate. When they grow, they reproduce "after their own kind," as though a kind of lock has been placed upon their cells and tissues.

Gregor Mendel: One of the first men to work with plants was a monk named Gregor Mendel. He ran a long series of experiments breeding pea plants in his garden. Mendel became famous, known as the Father of Genetics. His work still remains unchallenged.

Mendel discovered that each plant cell contains chromosomes, which control the heredity of the plant. Hereditary factors determine its size, shape, fruit, and so on. He named these factors genes. He found that the genes are passed along through successive generations, and allow for prediction of the characteristics of the plant. Mendel bred tall pea plants with other pea tall plants, thus developing new taller varieties of plants which could grow better in crowded gardens. In other words, by careful selection, he was able to modify the plants slightly, but they remained peas. Mendel attempted to modify the genes themselves, but the plants died immediately. He realized that this genetic code was very rigid and defied changes from outside forces. Darwin, on the other hand, had proposed that all living things would evolve. Mendel's work suggested he was wrong.

Mendel published his work shortly after Darwin brought out *Origins*, but it went largely unnoticed until after 1900.

DNA: In 1953, two scientists, Watson and Crick, discovered that chromosomes and genes were actually made up of long strands of DNA, a complex organic acid. All living things, including plants, contain DNA which records their genetic code. There are no exceptions. Therefore, every organism's characteristics are determined by its DNA code, which cannot be broken. Tampering with DNA causes it to die. This explains the findings of Mendel's experiments, but also suggests that random mutations, can never really take place.

Incredibly, both plants and animals seem to have an amazing capability built into its DNA code: adaptation. This means, simply stated, that they have been given the capacity to adjust to their surroundings, without making permanent changes in their inherent characteristics. Every organism can adapt. For instance, some trees lose their leaves in winter; this is adaptation. Other trees do not. In every case, the tree remains a tree, true to its species. Rabbits turn white in winter, brown in summer. Dogs and horses grow more hair in cold weather, then lose it as temperatures grown warmer. I could go on and on. We have all seen examples of adaptation, which is not the same as mutation. Yet Darwin seems to have overlapped the two in his writings. He proposed a term called permanent adaptation, in which an animal could permanently adjust its body. He did not know about genes and DNA, and was convinced that the boundaries within the species were flexible. Obviously, he was wrong. If Darwin were alive today, he would certainly be forced to withdraw portions of his Theory of Evolution, perhaps all of it.

There is now a vast store of evidence that refutes the mechanisms which Darwin proposed. I will devote several later chapters to discussions of genetics, mutations, and fossils.

So read on.

Chapter Six

WHERE ARE THOSE MISSING LINKS?

he study of fossils is an interesting one. Without a doubt, there are mineralized bones and even whole skeletons of ancient creatures embedded within the earth. The question is, when did they live? How did they die? There has been increasing interest in digging for fossils since the time of Darwin, resulting in an entirely new field of science called paleontology. These scientists descend upon new discoveries, remove the fossils, then attempt to catalog them. Much progress has been made in finding and digging fossils; with thousands uncovered since 1880.

Still, with all of this work, a striking fact remains: there are no missing links. You will recall, these were the remains of intermediary animals which should have occurred along the chain of the evolutionary process. Darwin himself expected that the ground would be full of them. He stated that complete evidence for his theories lay waiting for the eager shovels of his students. Much probing and digging has since been accomplished. Therefore, we must look at the concluding results:

1. Various fossils have been found at different levels in the earths crust. These often occur in groups, called beds.

2. Many fossils seem to represent extinct types of animals, without any clear relationship to those alive today.

3. A large number of fossils are of reptiles, showing little resemblance to warm-blooded mammals.

4. The ages of these fossils cannot be established with any accuracy, contrary to common opinion.

Please note the last two points above. The vast majority of fossils show skeletons of reptilian species, or in some cases, fish (piscine) species. Very few are birds; even fewer are mammals, that is, warm-blooded animals like dogs or cats. Very few fossils have been located which are traceable to mankind, or to any human ancestors.

Doesn't this strike you as odd? Surely, more advanced species such as mammals or humans would have been numerous. Mice, for instance, bear young at a fantastic rate, as do rabbits and other related animals. So where are the fossils? Where are the intermediate types which support evolution?

The oldest human fossil is dated up to 50,000 years old. The remains of most mammals are also very recent, dated in thousands of years, not millions. Deeper within the earth are beds of reptile and fish fossils, with no apparent link whatever to modern species. It is as though an entire group died out, later replaced by a new order of creatures. In between, there is a blank page in the fossil record. This is not what Darwin predicted, and does not support the theory of evolution.

Mammals, as we know them, appeared rather suddenly in the history of ecology. They were followed by the appearance of humans, without any evidence of a long chain of evolving ancestors. This is what the study of fossils tell us. The conclusion points to creation as the answer, rather than evolution.

Yet, many scientists prefer to ignore the evidence, apparently hoping for some big discovery to support them. Actually, the trend seems to be running against them, in terms of radio-carbon dating.

Carbon Dating: This is the method designed to place correct ages on ancient relics. When I attended college in the 1960's, carbon-dating was frequently cited as proof of dating for ancient finds. However, carbon-dating cannot be used on fossils, since they don't contain any carbon. Scientists would use samples of carbon in the soil where the fossils were found to place a date on the fossil. At best, this was an estimate, and, as it turned out, a poor one.

I won't go into the mechanism of radio-carbon dating fully, other than to say that it measures the decay of the Carbon-14 isotopes, which are faintly radio-active. This method is useful within a limit of about 30,000 years. Beyond that, it is inaccurate. This means that carbon dating of fossils, or even soil around fossils, is meaningless. They might as well throw darts at a dart board. The accuracy of the dart-method might, in fact, be better.

Recent findings of Dr. E. C. Pielou have proven the carbon-dating method to be absolutely useless for any artifact older than 30,000 years.

Yet scientists still continue to cling to facts supported by carbon-dating.

Cesium: The element Cesium was suggested to be a benchmark for dating fossils. Cesium is also an element which has isotopes, but its decay is much slower. The thinking was that Cesium, a metal, would remain lodged within fossils along with other minerals in the process of fossilization. This could then be tested for age by the relative strength of the radioactive signal of the Cesium isotope. In concept, this was a good theory. In actual fact, it does not work, since Cesium, in all its forms, is rather soluble in water and most fossils have been exposed to ground water for thousands of years, at the very least. Thus, Cesium testing of any fossil of a fish or amphibious reptile would automatically be invalid, because their contact with water would reduce the amount of Cesium.

Volcanic Ash: As a fall-back effort, a few scientists postulated that fossils could be dated if volcanic ash was present in the fossil bed. The idea was that certain volcanoes would give ash of specific composition. Again, much of this ash is water-soluble, and it proved unrecognizable after a relatively short period of time. In fact, actual ash samples from the famous eruption of Krakatoa are now significantly different in composition from the time of the blast in 1884. This idea has largely been discarded.

Accretions: Historically, archeologists have relied upon the concept of accretion, meaning the accumulation of soil and debris slowly built up in layers around an object. If an object or fossil is under 50 feet of soil and cover, the exploring team estimates the age of their find based on the depth of this covering layer. Theoretically, the accretion will build up at an average rate, of approximately one inch per 100 years. Thus, a fossil found at 50 feet can be dated as 50 feet x 12 inches/foot x 100 years/inch = 60,000 years. You get the idea, I'm sure.

It is easy to see the difficulty with this method; it is dependent upon winds and weather. Bury a child's toy in the sand on a windy day. How long does it remain buried? It can become uncovered in a matter of hours. There can really be no such thing as an average rate of accumulation. Accretion, as a scientific method, represents at best a wild guess at the age of any buried object. Fossils certainly cannot be dated with accuracy using such techniques. Surprisingly, this approach is still used.

Taking an overall look at our discussion, we can see that Darwin's theory looks shakier than ever. Fossils, after all, were supposed to provide the key proof for his concept of organic evolution. Although many fossils have been uncovered, they fail to prove his theory. In fact, they have increased the number of questions. Let's recap:

> 1. No orderly chain of fossils has been found. Where fossils are found, there are long breaks or empty layers in the soil between the beds, suggesting long periods of time separating the formation of the fossils
>
> 2. The fossil layers that do exist appear to be almost completely unrelated to subsequent layers.
>
> 3. It has been impossible to date accurately most fossil finds, so that any calculated ages are essentially guesswork.
>
> 4. Among man and the higher animals, no clear-cut missing links of any kind have been discovered in these fossil beds.

In actual fact, the formation of a fossil requires a peculiar set of circumstances. When a creature dies, its body will deteriorate as it remains exposed to the open air. You can test this assertion by observing the body of any dead animal along the road. You will notice the deterioration of the corpse each day. No fossils form along the side of a highway, nor do they form anywhere else in modern life. The bodies of such dead creatures must be either quickly covered or immediately frozen; otherwise, there is soon no specimen at all, let alone a fossil.

This explains the breaks or empty layers between the fossil beds. It is clear that ancient dead animals must have been covered with dust or ash, in order for their fossils to begin the process of mineralization. This set of conditions does not occur without some sort of great disturbance, possibly a global catastrophe, to provide the dust and ash. From this evidence, we also see that there obviously cannot be a steady progression of fossil remains. Darwin used a faulty assumption in predictions about the fossil record.

Another Look At Fossils

What do the fossils tell us, when we look carefully at the scope of the total findings? Darwin expected to uncover an orderly series of developing species, with one type of creature gradually blending into another. He expected to find a clear step-wise progression of evolving organisms. That is not the picture which emerges, however, based upon the recent summary of fossil-finds published by the National Geographic

Society. I have incorporated a chart taken from data in a recent article, *Extinctions*, published in the November 1989 issue of *National Geographic* magazine. Bear in mind, my chart is not a duplication of theirs; I have simply summarized their data here so that it is complete and easily readable, shown on page 47 as Figure 1. I would like the reader to study this chart carefully. Each line or bar on the graph represents one class of fossil going back in time. Several classes of fossils are represented there, including basic groups such as dinosaurs; the line shows the earliest age estimate for each group, as well as the most recent age being estimated.

Please note that there are absolutely no fossils existing for complex, multi-celled animals prior to a point roughly 600 million years ago. There may be fossilized remains older than this although such dates are only estimates but these invariably are the fossils of single-celled forms such as algae. Basically, 600 million years is the cut-off point; there were no animals living before that time. All complex creatures have arisen since that date, regardless of the assertion that our earth is at least 2 billion years of age. In other words, higher forms of life appeared rather late More importantly, the complex animals seemed to arrive all at once, not slowly over long periods of endless time. This fact has not been well received by many scientists. After all, many of them were raised on a diet of Darwinism and such a sudden appearance of higher life-forms strongly suggests an act of Creation. It is a hard pill to swallow.

Still, facts are facts. An increasing number of fossil specimens have been found which are dated at 530 million years. This seems to be the window of arrival for many creatures. The number of species grew fantastically at that time, a veritable explosion of new types not seen before. An excellent study, first published in 1993, has demonstrated that all of these animals seemed to suddenly appear on the earth within a span of 5 million years or less. The study was well-documented and compelling; even the magazine *Scientific American* made note of it. Obviously, our own conclusion must be that these species did not evolve, for they did not have enough time to do so.

Here we have the abrupt and unexplained appearance of hundreds of complex new animals, many of them related and/ or similar to animals living today. How is this fabulous event to be judged? Could we say that it happened by chance, in a random fashion? Would not the laws of probability make such a thing impossible?

These same conclusions were recently confirmed and reinforced by an article called *The Evolution of Life on the Earth,* written by Stephen Jay Gould, a professor at Harvard University. This article appeared in the October 1994 issue of *Scientific American*, and was intended as an explanation of how life on earth began. Remarkably, Professor Gould concluded that Darwin's revolution did not fully account for the existence of life. He said, "Darwinism is woefully incomplete in its grasp of the beginning of life." Obviously, I agree with Dr. Gould's statements.

Why was Dr. Gould so perplexed? The answer is rather simple, you see. All of life was created by a wise and loving God. Life arose very suddenly in terms of time. Within a short span of time, there were many new creatures, busy eating and exploring and reproducing. They soon filled the earth; but they did not evolve. This can easily be proven, since many of these fossilized life-forms are still alive today. Current specimens compare directly to their ancestors, cell for cell. They have not changed, except to produce additional similar varieties.

So then, what does the fossil record show? It demonstrates that most complex species arose rather late. Each new phylum (or major group) arose at different points in time, with no clear link to other earlier phyla. Each new group showed great diversity and surprising vigor, taking full advantage of the world around them. They multiplied and filled the earth.

The fossil record also shows that certain large groups, such as the dinosaurs, arose later and became dominant. This is evident in Figure one, as the dinosaurs appeared rather suddenly and died out abruptly about 65 million years ago. There is no tangible sign that the dinosaurs evolved; however, there is ever-increasing evidence that they were all killed in a cosmic disaster.

Mammals followed the dinosaurs. As the great lizards disappeared, warm-blooded creatures rose to prominence. Mammals are with us today in great numbers. Fossilized mammals do not comprise the majority of the fossil discoveries; their fossils are more recent, perhaps 100 million years or a bit older. Many mammals have become extinct within the last 10,000 years; their whole bodies have been found, frozen and intact, as a clear sign of another great disaster on the earth.

Mankind provides almost no fossils. He is the most recent and last of God's creations upon the earth. The shortage of human fossils is easy to explain. Man's history is measured in thousands, not in millions of years.

The fossil record really does not verify the process of organic evolution.

There are no missing links. Without this supporting data, evolution cannot be taken seriously.

Now read on.

The Fossil Record - Summary of Recent Data

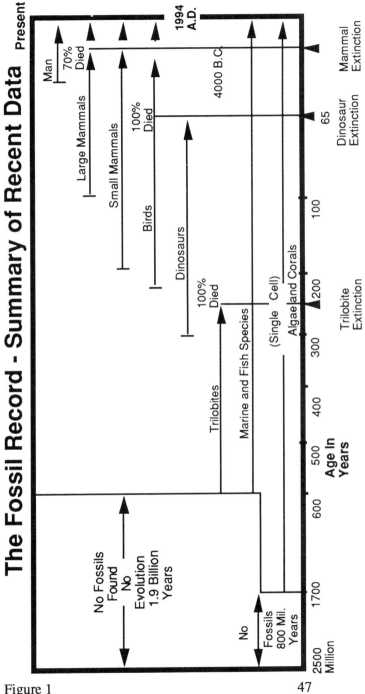

Figure 1

Chapter Seven

WHAT ABOUT MUTATIONS?

An understanding of mutation is basic to the entire concept of evolution. Darwin studied animals (and plants) which had apparently crossed the genetic boundaries with great ease. For example, if a species of mice lacked food, changes would gradually occur making them better able to compete for food. They might become faster, taller or smarter, maybe all three. This happened mainly through the catch-all methods of natural selection and mutation. Eventually, new generations of better, stronger mice would result. These mice were more fit to survive, and, over time, replaced the older versions. Eventually, a new strain of higher or improved creatures would result. This was, and still is, a cornerstone of Darwin's theory. Does it work? Please remember, Darwin never ran any controlled experiments; he only observed groups of animals while on the voyage of the Beagle.

Let's take the example used above, the common ordinary white mouse. Today, we have a great deal of information about these mice, since they are raised by the millions and used for laboratory experiments. We also now know enough about DNA and the genetic code that we can predict inheritance traits in laboratory mice with reasonable accuracy. There are a number of commercial lab-supply animal factories which raise and sell these mice to universities, etc. Such operations have become expert in assessing mice for use in experiments. Workers have noticed that, every once in a while, a mouse is born with a defect, such as an extra foot, resulting from a genetic flaw, or mutation. These mice can then be bred to reproduce this mutation in their offspring. Unfortunately, most of them die. In fact, Jackson Laboratories Inc., reports that only one defective mouse in 300 is a survivor; the other 299 will soon weaken and die. One example of a survivor might be a mouse with a longer tail. This is not an organic weakness, and the mouse leads a fairly normal life.

So let's look at this example more closely. Here is a case of an actual sturdy, mutant mouse, occurring very rarely, as Darwin might have suggested. It lives a normal life with a longer tail, even though the tail does nothing to make it more fit to survive. To achieve this one alteration requires 300 mutations, spanning several million mice over a period of 50 years. These are calculated facts, not theorizing. Therefore the success rate for mice mutations is 1/300 (a very low figure). Further, it requires 50 years to develop a survivor. Now we must add in the factor of fitness; not every survivor mouse will necessarily be more fit or capable. Obviously, only a tiny fraction of the survivors would qualify as improved. Let's say one out of two is stronger.

Now we are in a position to test Darwin's theory of evolution, with a set of concrete data. So let's run a little study of our own here, shown as MOUSE STUDY: Figure 2.

MOUSE STUDY

Item 1: Age of Earth = 2.5 billion years

Or   = 2500 million years

Item 2: Rate of Mutation for Mice =

1 survivor per 50 years

1 stronger per 100 years

Item 3: Success Ratio = 1/300

Calculation: Number of possible changes

or mutants in 2500 million years

A. For Survivor

$$\frac{2500 \text{ mil}}{50} \times \frac{1}{300} = 133,000$$

B. For Stronger

$$\frac{2500 \text{ mil}}{100} \times \frac{1}{300} = 66,500$$

What this study demonstrates is that the evolution of the white mouse must take place with no more than 133,000 mutations or genetic changes. To get a stronger mouse, only 66,500 genetic changes are available over the 2.5 billion year period. This is a very small number over a very long time span. It also assumes that the earth remains the same over that entire period; in other words, no catastrophes could occur to kill off the mice. Bear in mind, this is only an example, but it uses estimates which are founded upon actual numbers. This leads us to three obvious conclusions:

1. It is possible that this variety of white mouse could be altered very significantly over the course of 2.5 billion years. Starting with a large population of regular mice, the evolutionist would predict that there might be a population of improved mice after 2.5 billion years. This is possible, according to our figures.

2. However, it is considerably less likely that a single-celled amoeba could mutate by altering rapidly enough to become a mouse, even if it happened at a faster rate than in the example. Even millions of mutations would not be sufficient to improve from a single-celled being to a complex warm-blooded mouse, with all of its integrated body systems such as heart, lungs, digestion, reproduction, etc. Please remember: the time limit is now fixed at 2.5 billion years. So evolution would be

required to happen inside that time-frame.

3. Let's move on to consider the entire scope of creation, both plants and animals. With all the thousands of species and ten-thousands of varieties, how many billions of mutations would be required to develop our entire ecosystem spontaneously? Would it take 10 billion mutations? 100 billion? Would there ever be enough time? Could anything mutate by itself that rapidly, and without any direction? Can we reasonably believe that such a scheme of evolution is possible within the short span of 2.5 billion years? It certainly looks doubtful.

This example has demonstrated for us the stern mathematical principles known as the Laws of Probability. Even one simple mutant change in a single species, such as the white mouse, would require thousands of genetic modifications, spread across millions of years. In the meantime, there is the possibility of external factors having an impact on the survival, such as a large population of predators. Our example did not even take into account such a possibility, but it's conceivable that any number of factors could play a significant impact on the final outcome of the process. Darwin skirted this problem by saying that the mouse could prevail over endless time. As we have shown, however, only a limited time was available.

The logical conclusion is that mutation is far too slow a process to account for the development of all the present-day species on the earth.

The fact that mutation was supposedly random and undirected makes the probability of evolution become essentially zero.

The Fruit-Fly Saga

Even with the above arguments, there are those who argue that the mechanism of mutation can explain the development of plants and animals. They say that, yes, it's a long-shot, but evolution is theoretically possible by mutating creatures into new forms. This line of reasoning might convince some people of evolution, or at least to wonder about it.

So we need to get down to hard cases. Does mutation really take place or not? That really is the question. Darwin merely theorized about it; he never actually tried this mechanism, nor did he prove that it can advance any species. As mentioned before, Darwin never did laboratory experiments of any sort. The reader should keep that in mind.

I will now describe for you an experience from my own past. I call it the Fruit-Fly Saga. It was a practical answer to the question of mutation, learned firsthand by me at the age of 23. At the time, I was taking a college course called Genetics

311, which was the study of genes and how they control heredity in organisms. Included in this class was a laboratory experiment in which we bred a colony of fruit flies, then rated them for various inherited traits. For instance, some flies had transparent wings, while others had spotted wings. These traits were passed from parent to offspring, and could be traced through several generations. Fruit flies breed quickly, with a new generation appearing about every ten days, as I recall. We soon had many generations of these bugs, kept in jars, labeled according to their characteristics.

The purpose of our experiment was to be a study of mutant fruit flies. The problem was, however, that none of them had mutated on their own, even though there were thousands of them. They all reproduced 'in kind,' you see. So we forced them to mutate by using various chemicals. We exposed their eggs (which they laid on the surface of rotten banana-peels) to a variety of toxic gases: nitric oxide, sulfur dioxide, ozone, trichlorethane, and others. Most of the chemicals killed the eggs. The fruit flies survived the nitric oxide, however, and the eggs hatched.

Eureka! We found that about one fly out of 50 had a mutation. We examined these mutants, noting that some lacked a wing, others a leg; some were apparently blind and kept bumping into the side of the jar. We carefully separated all the mutants and placed them into a jar especially prepared with juicy, rotten banana skins which we called Fruit-fly Heaven. Then we waited for them to reproduce. To our surprise, they did not have any offspring. You see, they were completely sterile.

In vain, we students hunted for a larger, stronger fruit-fly. Darwin assured us that there would be one, so we dutifully searched among the offspring which came from the gassed eggs. We found no strong, virile new variety of fruit-fly, only weaker flies. Our experiment did not bring a fitter type of insect into the world. In fact, all the mutants died one by one, leaving no survivors. I want to repeat this: there were no survivors! Even though we had prepared an ideal place for them and they were protected from predators or catastrophes, they all died off, because they were weaker.

We learned that men cannot tamper with the genetic code of fruit-flies. Change their genes and they die. The reason is the complex nature of DNA which contains the actual genetic file for each living creature. This DNA code is fixed for each species, and dictates its specific traits. To summarize: If you tamper with the DNA, the creature will either die or be greatly weakened; those that survive are usually unable to reproduce.

I will have more to say about DNA and the gene-code in the next chapter.

Evolution vs. Adaptation

Charles Darwin didn't know enough about genetics to differentiate between evolution and the process of adaptation, a characteristic which all creatures possess. Every plant and animal has a God-given capacity to adapt to its surroundings. Examples of this abound in nature. We are aware that wolves in the far north have white fur in winter, and darker fur in summer; such changes are adaptation. Animals are able to accommodate changes in climate and surroundings, but they do not change into another species. Whether in winter or summer, the wolf retains all the characteristics of a wolf. Let me clearly restate the principle:

Adaptation does not lead to Evolution.

Further, we can see from Mendel's work, that men can selectively breed one species, such as corn plants, to be very tall or resistant to cold temperatures. Such selective breeding produces what are known as hybrids, that is, a special variety with desirable characteristics. In no way does the hybrid corn represent a new species; it is still corn as we know it. It's too bad that Darwin didn't see this in his day. His concept of mutation has been thoroughly discredited by modern experimentation. Mutation does not lead to evolution; in fact, no known mechanism produces evolution.

The purpose of this chapter has been to discuss the Darwinian view of mutation as it relates to evolution. Essentially, the statistical probability of mutation occurring and producing a new creature is zero.

When we look at actual case studies, it's obvious that life did not arise by pure chance. Also, it is easy to see that, once various forms of life were present on earth, they did not mutate into new or different species. When you realize that evolution, if it had taken place, would have had to happen within the span of 2.5 billion years, then you can see how impossible it is that the thousands of current species developed in this manner.

Such calculations as I have shown here are easily understood by students of first-year algebra. So why do college professors persist in teaching evolution?

Read on.

Chapter Eight

MEN AND MONKEYS

id mankind actually descend from monkeys? There are certainly some similarities between men and the other primates. These similarities are often used as proof of common ancestry. Scientists have always been fascinated by the entire primate family, the group which includes apes and monkeys, because they possess high intelligence and show many human-like emotions. Numerous studies have been carried out to show how closely these animals parallel our own behavior.

While very interesting, it is not proof of evolution. Not at all. Apes remain animals, with certain inborn characteristics. Humans are significantly different, which becomes obvious when you observe closely the apes' behaviors. Men can speak, reason, and organize information in written form. No animal can do this. Parent apes do not pass along learned information to their offspring. Men, on the other hand, can provide a full set of information to their children. Apes live on a basic level; men are capable of higher complex thinking.

The key here is the idea of inborn capability. Men are born and almost immediately begin to try to speak. An ape is born and tries to get up. Each one has a different set of drives and instincts. Humans want to communicate, while animals instinctively desire to move.

Where do these differences come from? Clearly, it is genetic in its origin. So, why not simply cross a human with an ape, to produce a creature with the traits of both? Easier said than done. As we mentioned earlier, Mendel found a lock on the gene-code which cannot be breached. This code is locked inside a long spiral of DNA, which looks very much like a zipper on a coat. These DNA spirals actually have two matching ribbons, with tiny links which join them like rungs of a ladder. This is shown in Figure 3. Every species has its own distinct DNA combination, which is very specific and rigid. The father and mother each contribute half of the code (one side of the zipper) in their offspring, and this determines all of its characteristics for all time. DNA is organized into chromosomes, threadlike structures inside each cell of the creature's body. Each species has its own number of chromosomes which vary greatly in length. The DNA code is extremely complicated. For humans, each DNA ribbon may have from 20,000 to 100,000 links, each link bearing a specific piece of information. Collectively, they determine the entire size and function of the host creature.

Sound complex? Well, it is. The DNA code is so difficult that recent experiments for mapping the human gene-code by computer are only about 35% completed. Scientists have been working on it for seven years already.

Once we realize the superb complexity of the DNA, several other thoughts immediately come to mind:

1. Could anything as intricate as DNA simply evolve by itself, without help?

2. Could DNA alone (excluding all other considerations) have evolved by selection and mutation in the short span of 2.5 billion years?

3. By what mechanism would the DNA of one species alter itself, so that another species emerged.

This last question is the key to our discussion in this chapter. Man could not be related to the apes unless they had an identical set of chromosomes. Humans have 23 pairs of chromosomes within the DNA molecule. Monkeys, gorillas, and apes have 24 pairs. This difference makes cross-breeding totally impossible. If man evolved from apes, he somehow would have had to shed that extra pair. (Goodness knows how many hundreds of billions of mutations would be needed to do this!) On the other hand, if man had evolved from a common ancestor with 23 pairs, then where did the apes get 24?

Further, if we refer to Figure 3, you will see that one

DNA strip requires an exact matching set to form the complete DNA code or ladder. In humans, this length runs to a total of 100,000; in apes, these strips are somewhat shorter. Thus, even if we could eliminate the problem of the extra pair of chromosomes, we still have to deal with matching the different lengths of DNA. You can easily see the magnitude of this puzzle, with its limitless variations.

Based on this evidence, it is safe to say that man did not evolve from apes. So far, no one has found a common ancestor.

Read on.

DNA - Model

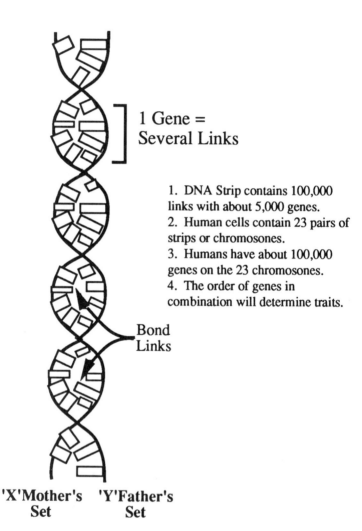

1 Gene =
Several Links

1. DNA Strip contains 100,000 links with about 5,000 genes.
2. Human cells contain 23 pairs of strips or chromosones.
3. Humans have about 100,000 genes on the 23 chromosones.
4. The order of genes in combination will determine traits.

Bond Links

'X'Mother's   'Y'Father's
    Set   Set

Figure 3

Chapter Nine

CHAOS AND ENTROPY

harles Darwin never studied chemistry. Even if he had, the field of chemistry has since advanced far beyond his wildest imagination. In the interest of keeping this book simple and readable, I will not discuss those various aspects of chemistry which refute Darwin's theory. However, since chemistry is my own field of interest, I can easily see that Charles Darwin knew very little about it.

For instance, the DNA which we discussed in the last chapter is really a very complex organic chemical: deoxyribonucleic acid. It is manufactured inside living cells, and is so complex that men have been unable to duplicate it. When DNA is removed from the cell, it tends to deteriorate rather rapidly. In other words, DNA survives and grows only within a living cell. This principle is important, since all living things exhibit very complicated body chemistry. Very few of these chemicals can survive outside the organism. Two such examples are blood and urine, both of which must be refrigerated to keep for laboratory analysis.

Let's restate this principle: All complex organic chemicals tend to degrade rapidly into simpler forms. The exception to this rule seems to be what occurs inside living cells and tissues, where simple compounds (such as sugar, starch, and water) are built up to become more complicated molecules and tissues.

There is a name for the above principle of chemical degradation: Entropy. This principle is so well-proven that it is considered a Law. It is officially called the Second Law of Thermodynamics which states that any complex substance will always tend to degrade toward simpler forms.

Consider a complex organic material such as common white sugar. You all know what happens when sugar gets wet. It dissolves. Once in solution, it is attacked by a variety of forces and unbalanced, changing it into a series of ever-simpler forms. These include alcohol, aldehydes, acetic acid and carbon dioxide. Within a few days there is no sugar left, only a mixture of simple end-products. This is a common example of Entropy at work.

There are two laws of thermodynamics, as follows:

I. Enthalpy: The total amount of matter and energy in the Universe is constant at any given moment, with nothing lost and nothing gained. Matter and energy are in perfect equilibrium at all times.

II. Entropy: All complex substances tend to degrade toward simpler compounds, with the attendant release of energy.

At first, the two laws may seem to contradict each other. Actually, there is no contradiction. The first law deals with large systems, such as an entire galaxy; the second law deals with small-scale situations, such as one set of conditions on one planet.

Which law applies to Darwin's theory of evolution? Quite clearly, it is the second law, because evolution supposedly took place here on the earth under an ideal set of circumstances. Evolution would be considered a small-scale situation, since it involves one original single-celled animal arising out of the muck. Darwin never heard this Second Law of Thermodynamics (it was developed after 1900). It completely negates his primary thinking in that simple compounds do not tend to become more complex, not without a lot of help. If left alone, that first single-celled creature would have died and quickly decayed. In no case would it have become more complex. The only way to get around this basic law is to put energy into the change. In plain language, evolution would have needed intelligent assistance. Every experienced engineer knows that complex things cannot be fabricated without intelligent assistance. A clock will not assemble itself by random chance; energy has to be involved. A fiber of nylon cannot be developed unless energy is put into the reaction. Examples of this principle are almost without number.

There is a universal tendency in our world for orderly things to become disorderly. We see many situations which drift from reasonable order into pure chaos. The outcome of the Second Law is always disorder; the end-result of entropy is chaos. Far from proving the possibility of random evolution, the Second Law strongly proves that it could not have happened.

Our world was organized by a Supreme Intellect, who with his own energy overcame the effects of chaos. Creation could not have been a random event. It was planned.

More about this Great Plan later.

Read On.

Chapter Ten

OUR YOUNG EARTH

s I mentioned in Chapters 3 and 4, there has been an ongoing debate about the Bible story of creation. According to the Darwinian argument, the earth is very old, an ancient fixture in the timeless heavens. It could not have had its beginning in recent times, for its crust and mantle suggest that it is crumbling with age. Further, the Bible account is not scientific, containing odd mythical references to good and evil. In addition to this, the presence of fossils suggests that the earth had an on-going population of various animals living on it over long periods of time.

We have all heard these arguments before, and, in fact, they do contain some points that are true, as follows:

1. The Bible was not written to be used as a scientific text book; it was intended to state God's purpose and plan for mankind.

2. The Bible does deal with spiritual and moral questions, which scientists usually avoid.

67

3. The earth does have fossilized remains of past generation creatures, many which are now extinct. The Bible may refer to these, but only indirectly.

As I lay out these points for your consideration, you may know of Christians who argue heatedly that the earth was created in six days of 24 hours each. We have already dealt with the questions of relativity and time-flexibility, as it relates to the Bible's statement regarding the creation of the solar system on the Fourth Day. Frankly, until completion of the Fourth Day of Creation, there was no certainty that a Biblical day was 24 hours in length. Regardless of how long these creation periods were, there is a great deal of proof that our earth is not very old.

The idea of a young earth would have been laughable fifty years ago. Today, it cannot be laughed off, for many discoveries have been made through the advent of moon shots and space shuttles. The earth has been studied and measured as never before, beginning in 1957 with the orbit of the Russian Sputnik satellite The advance of space travel provided an accurate measurement of the earth's rotational speed, the tilt on its axis, and the strength of its magnetic field. With the moon-shots came more precise testing of the lunar orbit and surface features. The resulting data was so surprising that it was retested repeatedly, by both Russian and American astronauts. Finally, when the facts could no longer be disputed, they were published. I have listed them below in order:

Earth factors

A. Rotation: The earth is spinning like a top, as we know. At the equator, its speed is about 1,000 miles per hour. This speed is sufficient to hold the planet in a stable position on its axis, but not so fast that the water in the oceans gathers at the equator in a bulge. Darwin assumed it had always done this, through eons of time. Not so. The earth is slowing down, gradually losing speed at a steady rate. Within several thousand years, it may lose enough speed to become unstable and begin to wobble on its axis.

It may be more interesting to work backwards. Looking back to less than one billion years ago, the earth would have been spinning so fast that all the water would have gathered at the equator and been flung off into space. The land continents would have slipped over the mantle beneath and bunched together along the equator like a mound of corn flakes. It would have been a catastrophe like none we could imagine. So much for the idea of our planet being one billion years old.

Let's look back just 50,000 years. At that time, the rotation of earth would have been enough to move the oceans toward the equator, increasing their depth by 200-300 feet. The poles would have contained no water at all, and the ocean beds in the far north and south would have been mostly dry. Central America would have been essentially submerged, with only the mountain ranges protruding as islands. Large portions of North and South America would have been under water. Florida and Louisiana would have vanished beneath the waves. Much of Africa would have been covered with water as well. This condition would have been more or less permanent. It would most assuredly have affected the ongoing process of evolution by disrupting the habitats of the animals as well as their breeding. Where rain-forests grow today, mighty seas would have rolled. There is no evidence of that ever happening.

Let's face it: if the earth ever were spinning that fast, we would see clear signs of great geophysical upheaval. Any buckling crust or surging sea leaves its marks etched in the land. There are no such marks today. (The few exceptions are the signs of Noah's flood.)

The earth may be losing speed, but it has not been in existence long enough to reveal effects of earlier rapid rotation.

B. Tilt of the earth: During the 1960's, astronomers found that the earth was showing a slight shift on its

axis. When compared with earlier data, the earth
appeared to have recovered from this declination (or
tilt). Currently, the center line of the axis of rotation is
tipped 23° off the vertical. The recovery seemed to be
about 3.° In other words, the earth tipped over as far
as 26° at one time. When the rate of recovery was
figured, it appeared that 6,000 years had elapsed since
the earth was tilted to 26.° That means that it recovered
1° every 2,000 years or so.

Even more surprising, it was discovered by a
Russian named Milankovich that the earth was showing
some wobble while spinning on its axis. Yet, in theory,
the earth should be perfectly stable. This information
suggested that, in times past, the earth had been forced
over onto its side, much like a spinning gyro that is
bumped with the hand; it continues to spin and
eventually rights itself, but in so doing it tends to
wobble on its axis.

Was the earth bumped? Evidence suggests that
it was, or, more correctly, it was the victim of a near-
miss. If such is the case, our planet has not always
been at a tilt of 23°. The axis of earth has changed very
significantly, some of it during recent historical times.

The earth is not that ancient fixture hanging in
the timeless heavens. It has moved, and not so very
long ago.

C. Magnetic Field: The magnetic compass has been in use for about 400 years. All ships and airplanes depend on it for direction. A compass, in turn, is guided by the earth's magnetic field, always pointing towards the north. When I was a Boy Scout, we were taught that true north was slightly different from magnetic north. On the map, the North Pole was not the place to which the compass pointed; the needle always points toward magnetic north. So there are really two poles; the magnetic pole is at a different location from the one at the center of the earth's rotation. In fact, they are more than 900 miles apart, although they should be the same. The rotational poles have apparently moved from their original positions by 900 miles, although not within our recorded history.

When did these poles shift? This date is possible to estimate, since the earth's magnetic field is still functioning and can be used as a measuring device. This field is exactly like that of an ordinary magnet, consisting of two poles with lines of flux, or electrical current, flowing between them. Over the past 100 years, the strength of the earth's field has been measured at different times. Darwin knew of the earth's magnetic character, but he assumed it remained unchanged. Not true! The strength of the field is diminishing at a steady rate, losing half of its strength every 1,400 years. To say it another way, the field has a half-life of 1,400 years. We would compute as follows:

1.1,400 years ago, field was twice as strong as today.

2. 8,400 years ago, it was 64 times its current strength.

Thus, 8,400 years ago (in 6406 B.C.) the earth's field would have been powerful enough to cause spontaneous lightning bolts and sparkling displays across the sky. If we go back 11,200 years, the earth would have had the magnetic force of a large star, disturbing our moon and solar system. Twenty-thousand years ago, the electrical current flowing through the planet would have melted the rocks within a 100 mile radius of each pole.

Is there geological record that any of this occurred? To date, no such evidence has been found. Frankly, it would appear from the illustration above that our earth is well under 10,000 years of age, and probably not more than 6000 years, based on its magnetic field.

II. Moon Factors: The Moon is about 240,000 miles from the earth and revolves around it once every twenty-nine and one half days. This distance from earth is rather ideal; if it were less than 200,000 miles away, our oceans would experience enormous tides that would rise nearly 150 feet every day! If it were further away, it would easily escape the gravity of earth, spinning off into space.

A. Recession: Yet in the last 60 years, scientists have found that the moon is receding from earth. With every orbit around our planet, it pulls a little further away, increasing its distance from the earth. The moon's orbital radius is increasing about three miles each year, meaning that in 333 years the moon will be at a distance of 241,000 miles. Again, working backwards, we would note that in the year 1661 A.D., the moon stood 239,000 miles away; during the life of Jesus, it was only 234,100 miles and in the days of Abraham, the moon stood even closer, only 228,200 miles. You can see the progression easily. What about that problem I mentioned in the first paragraph? When would the earth and the moon have been less than 200,000 miles apart? That is easy to calculate:

$$\text{Time} = \frac{\text{Difference in miles} \times 333}{1,000}$$

$$\text{Time} = \frac{(240,000\text{-}200,000) \times 333}{1,000}$$

$$\text{Time} = \frac{40,000 \times 333}{1,000} = 13,320 \text{ years}$$

This estimate tells us that roughly 13,000 years ago, the moon would have stood a mere 200,000 miles away. It would have appeared much larger in the sky and about three times as bright. Its gravity pull would have caused terrifying upheavals on the earth, in the form of tremendous tidal waves. Great earthquakes would have been common; volcanoes would have erupted with frightening frequency. Again, there is no evidence that such disasters occurred within the recent past.

Therefore, we must conclude that our current earth is younger than 13,000 years by a wide margin. Facts are facts.

B. Moon Dust: The moon collects dust from space. Its gravity acts like a vacuum cleaner, sucking in many tons of space debris continually. Unlike Mother Earth, the moon has no atmosphere; therefore, such particles do not burn up as meteorites, but fall to the moon's surface. In four billion years, scientists calculate that the moon could have gathered up to 160 feet of loose powder as a covering. This was a major concern in the mid-1960's, as the U.S.prepared to send men to the moon. What if the lander vehicle dropped into a soft, loose blanket of dust and became engulfed, never to rise again?NASA cautiously sent a moon-probe to determine the depth of the dust. What a shock! There was dust everywhere, as expected, but nowhere was it deeper than one inch. In most places, it was less than one-half inch in depth.

Using these figures at which dust accumulates

on the moon, scientists calculated the age of the moon to be about 8,000 years. There you have it! Our moon may have rotated someplace else previously, but it seems to have joined the earth's orbit about the year 6,000 B.C.

This happened on the fourth day of creation, just as the Bible had specified.

III. Geological Factors: The geology of North America shows many features that are relatively new or recent. The mountain ranges in the west are tall and impressive; the dry lake beds are full of salt, the residue of evaporating sea water. Much of this continent shows evidence of great changes having occurred in relatively recent times. How recent were those changes? Perhaps we can calculate an answer to this question below.

A. The Niagara Gorge is familiar to most of us, an enormous gash in the surface of the land. Most people who visit Niagara, look at the falls and the whirlpool. Few visitors pay attention to the gorge itself, which actually exists in two parts: the Lower Gorge (below the whirlpool) which was cut through soft land on a lower level, and the Upper Gorge which was cut through solid rock. Four of the upper Great Lakes pour their contents over Niagara Falls, which drops 180 feet over two sets of cataracts: the smaller American Falls and the larger Canadian Falls. It was not always this way; not long ago, there was only one falls, with all the water going over it. This caused erosion of the underlying rocks, which eventually caved in and fell, and were washed down the gorge.

Men have been visiting Niagara since the early 1700's, a span of more than 250 years. Fortunately, some of those early visitors marked the point of the brink with measurements. There is a documented location of the falls in 1764. Thus, it is easy to determine how far the falls have moved by cutting away the rocks. Currently, that distance is just slightly more than 1,000 feet, or an average of 4.3 feet per year. This is the rate of erosion.

With that, it is easy to calculate the age of Niagara Gorge, at least down to the whirlpool, which lies about 19,000 feet below the current falls.

$$\text{Time} = \frac{\text{Distance}}{\text{Rate}} = \frac{19{,}000 \text{ feet}}{4.3 \text{ ft/year}} = 4{,}419 \text{ years}$$

Since the type of rock is basically the same over this distance, we can be quite sure that the Upper Gorge is around 4,500 years old. Erosion of the basin which forms the whirlpool added 500 years while the Lower Gorge took another 500. The age of the entire Niagara waterway is therefore 5500 years, not millions.

So much for Darwin and his eons of slow progress.

B. The Mississippi Delta is known to us in story and song, a huge alluvial deposit at the mouth of the Mississippi River near New Orleans. The delta is growing constantly, as the river dumps new deposits each day. In fact, the U.S. Army Corps of Engineers has ongoing data on this steady growth. They have kept these figures since the 1880's, so their data pool is very large indeed. While the silt deposits vary from month to month, the overall yearly rate of deposition is surprisingly stable.

Now then, it is a simple matter for these engineers to calculate the size of the entire delta area, which they have done. The weight of the entire delta can be figured in tons, although the actual number is astronomical. Once again, it takes simple mathematics to divide the mass of the delta by the yearly rate of deposit. This has been done various times over the last 50 years. The calculated age of the Mississippi Delta is 5,000 to 5,500 years.

It does not take great genius to notice that the age of these two geological landmarks is about the same !

What does this tell us about North America?

Clearly, our continent is young. It is consistent with the ages we calculated in our other earlier examples. There are further examples which could be added; the age of Lake Huron, which is estimated at between 4,000 and 5,000 years; the age of the Great Salt Lake, which seems to be around 4,500 years of age; the salt content of Mono Lake in California, which places its age in the range of 5,000 years.

Why bother to go on? The present earth is young, by any standards. It did not evolve, as Darwin claimed.

Neither did mankind.

Read on.

A Summary

Before moving on to the last three chapters, I want to recap where we have gone so far. It was my stated intention in chapter one to show that Darwin's theory of evolution is no longer valid. There is a great deal of modern evidence which refutes Darwin, and I have carefully laid out much of that data for you.

We examined Darwin's work and the basis for his theory of evolution, stating the main assumptions which underlie his thinking:

1. A timeless, endless Universe.

2. An ancient planet (earth) which developed an ideal

environment.

3. Given the above, a form of life had arisen spontaneously and evolved according to random occurrence.

I then set out to destroy these assumptions, using discoveries which have taken place since Darwin died.

We showed in Chapter 3 that the Universe is not ageless; it has an age which can be calculated.

Further, we have shown that the Earth is younger than the Universe, and has been inhabitable for 2,500 million years or less. It is not timeless at all.

We have shown with simple illustrations that life could not possibly evolve in this time span. Furthermore, if the earth proves to be still younger (which it will), the probability of evolution becomes even less.

We also noted that the Big Bang Theory follows the same scenario as Genesis One, except that the Big Bang gives no credit to God.

Further, we have pointed out that time changes near the speed of light. If something happens fast enough, time becomes flexible. This effect is called time dilation.

Next, we outlined the mechanism of chromosomes and DNA, facts unknown to Darwin. Each species has its own rigid genetic code, not easily broken. Each creature reproduces "after it's own kind."

We discussed fossils. Millions have been found, but none prove the existence of intermediate ancestors called *missing links*. Their absence is a serious blow to Darwin.

Natural selection and mutation were both explored. One concrete example of a white mouse was used to show how unlikely it is for one species to evolve significantly during 2500 million years, let alone all of the species which are currently alive.

We examined apes and monkeys, which are thought to be related to man.

We argued that man is very different, unique among the primates, with DNA that does not match that of apes. Men and apes are not cousins. In fact, men and apes are not related.

We discussed the Laws of Thermodynamics, in particular the Second Law, along with the principle of entropy or degradation. We suggested that evolution could not take place because the Second Law would prevent it.

These Laws were discovered 40 years after Darwin wrote his book.

Finally we talked about recent discoveries which show the earth to be young, not ancient. Recent geophysical findings place the age of the earth in the range of 5,000 to 12,000 years. These are only estimates; the exact age of our current world has yet to be accurately set.

Clearly, the weight of all of these arguments refutes

Charles Darwin and his theory of evolution. If Darwin were alive today, he would probably change his position or at least modify his views. Yet many continue to follow Darwin in the name of science, choosing to disregard the facts which have surfaced. Their choice to follow Darwin against the evidence is a step of faith rather than reason.

Darwinism has become a religious cult with its own doctrines and philosophy.

Worse, perhaps Darwinism has become a society of the ignorant.

The remainder of this book will be devoted to examination of two basic subjects:

1. Mass extinctions of the Woolly Mammoth and the Dire Wolf.

2. God's Master Plan for Mankind.

Unlike the first ten chapters, I will use the Holy Bible as a reference for this section, including the use of footnotes and a scripture index.

Keep your Bible close at hand. . . .And read on.

Chapter Eleven

EXTINCTION OF THE FITTEST

Preview of Time

In the beginning, God created the heavens and the earth.

At first glance, it would appear that God did the whole job alone, but such was not the case. There are other references to creation given in other parts of the Bible, which suggest that God was not alone. Look at the Book of John, Chapter 1, verses 1 through 3:

1. *In the beginning was the Word, and the Word was with God, and the Word was God.*

2. *The same was in the beginning with God.*

3. *All things were made by Him; and without Him was not anything made that was made.*

The *Word* here refers to Christ who was with God and helped in the creation of all things. In fact, Christ was the major architect in the creation plan. Furthermore, all things were created to be ruled by Him, as noted in Colossians 1:15-17.

15. *(Christ), who is the image of the invisible God, the first-born of every creature:*

16. *For by him were all things created, that are in heaven and on the earth, both visible and invisible, whether they be thrones or dominions or principalities or powers: all things were created by Him, and for Him.*

17. *And He is before all things: and by Him all things consist.*

It is apparent that God created our entire world in a joint effort with Christ. This included things that were visible as well as invisible, that is, a kingdom of spirits. Clearly, this refers to angels, who are invisible and have great powers. They were among God's earliest creatures. The creation of our earth took place after the initial creation of the universe and also the angels. Christ, himself, was present for all of this, as we read in Hebrews 1:2-4.

2. *(God) hath in these last days spoken to us by his Son (the Word) whom he hath appointed heir of all things, and by whom he also made the worlds.*

3. *Who being the brightness of his glory and the express image of his person, and upholding all things by the word of his power sat down on the right hand of the Majesty on high.*

4. *Being made so much better than the angels, as he hath by inheritance obtained a more excellent name than they.*

So, there we have it plainly stated that Christ, who made the universe, is the heir to rule the universe. Christ upholds all things by the word of his power, and he is of higher authority than the angels. We read later in Hebrews that angels are powerful ministering spirits sent forth by God. Our physical world (the visible created things) is made of particles of pure energy, compacted into small units called atoms which are dense enough to reflect light; therefore they are visible. In fact, all solid matter is made up of atoms, those tiny building blocks of the universe. Yet their internal energy is controlled by the word of Christ's power, which keeps them locked into their pattern so to appear solid to our eyes.

The visible worlds were created by God for a purpose: so in the fullness of time, they might all operate in harmony under the leadership of Christ, and, in so doing, bring glory and honor to God the Father, as noted in Colossians 1:10.

10. *That in the dispensation of the fullness of times, he might gather together in one, all things in Christ; both those things which are in heaven and on earth, even in him.*

God's plan was, and still is, to bring all things together

under one head, one leader, one Prince. Christ is that leader and Prince. He has total authority above all of the angels, and, when the appointed amount of time has passed, Christ will assume this role which was given to him by God.

Does our planet Earth have a purpose? If so, what is the reason for creating mankind? After all, there were already angels created by God before time began as we know it. They should have been able to run things just fine. Angels are very powerful and able to move about, unfettered by real bodies. Obviously, they do not consist of atoms which reflect light; therefore, they are not visible. Angels are free to zip through space, probably faster than the speed of light. So why not let them manage the earth and universe? The reason has to do with a serious war fought in heaven: Ezekiel 28:12-17.

12. You (Lucifer) were once an example of perfection. How wise and handsome you were!

13. You lived in Eden, the Garden of God, and wore gems of every kind. You had ornaments of gold; they were made for you on the day you were created.

14. You lived on my holy mountain and walked among sparkling gems.

15. Your conduct was perfect from the day you were created until you began to do evil.

16. So I forced you to leave my mountain. Because of this, I hurled you to the earth and left you as a warning to others.

Lucifer was probably an archangel, a being of high station and great power. He was greatly honored from the day he was created. Note the use of the term day here, for the angels were made very early, probably before the current universe. Lucifer existed in opulence and prestige, living on God's own mountain until he began to do evil. Then he was hurled out of heaven and down to the earth. This probably meant that God forced him to exist in our physical universe, stripped of his honors and authority. But Lucifer still had some of his power and continued to cause mischief. At some point, he became involved with God's physical creation and began to pollute it. Lucifer began to tamper with the creatures and to pervert them in some way. How do we know this? If you look carefully at the story of Eve and the serpent (in Genesis 3), you will note that this animal was already in alliance with Lucifer (or Satan), and willingly spoke lies to the woman. This is a key point. Eve was approached by an animal that was already cunning in presenting Satan's questions to her. How could the serpent have reached such a point of sin so quickly, unless it had become perverted? Think about that for a moment; we will return to that point later. Satan (Lucifer) actually began his evil work on certain animals, one of which was the serpent. Some of God's creatures had already fallen under Satan's influence by the time Eve was approached. Adam and Eve were created sinless, a special pair of intelligent beings whom God planned to use to combat Satan's evil. At that point, there was no general curse upon the ground. Creation was harmonious, or at least that portion which lay inside the Garden of Eden. Perhaps Lucifer held sway outside the Garden or perhaps not. The Bible is silent on that specific point. We do know that he had already been banished from heaven and spent a lot of time traveling to and fro upon the earth, as described in the Book of Job, Chapter 1:6-10. Cast out of God's presence, Lucifer had come to earth to make trouble; he brought one-third of the angels with him, and they spread a pall of evil across the universe. Perhaps this was the point at which chaos (or entropy) entered our existence. If so, Satan himself would be the author of the Second Law of Thermodynamics (see Chapter 9).

Let's examine this entire situation more closely: God creates the universe with the direct help of Christ (the Word). They have it functioning harmoniously. Then Lucifer rebels and is thrown out of heaven; he takes refuge, along with his angel followers, in our physical universe. They notice that God has given special attention to a small planet, Earth. It is a beautiful place, and life is already rooted there. Lucifer realizes that something unusual is taking place: another form of life is in existence besides the angels. He becomes concerned, and goes to investigate personally. What a shock! It is obvious that God has been working on something new! This could be a disaster for the devil and his henchmen, but they are unable to destroy the earth outright. Lucifer no longer has the authority to do so. What to do? In his usual subtle fashion, he plans a form of indirect attack. He will subvert God's creation, bringing them slowly under his own evil control. He starts to work on some of the simpler creatures and finds success with the serpent family. He gradually gains control of other creatures as well, bringing about confusion, violence and chaos. The earth becomes a fearful place, with ferocious beasts battling one another, leading to the most dreadful state of affairs: survival of the fittest.

A Quick Recap

Let's recall some of the things we learned in the first ten chapters.

We learned that our Universe is finite and expanding; it began at one single moment in time, no more than 11 billion years ago.

We learned that the Earth is younger than the Universe, no more than four billion years old.

We learned that life has existed on Earth for only half that time, no more than 2.5 billion years.

We learned that the Earth has been in its present orbit for 12,000 years or less, and the moon for less than 10,000 years.

We also noted that our Earth shows signs of marked change, such as shifting poles and changing tilt.

We saw that North America is no older than 5,000 - 5,500 years.

Given these facts, what can we, as Christians, make of all this information? What does the Bible say about it? How can we reconcile all the different times and apparent disasters? Was the Earth destroyed at some point? If so, what were the circumstances? If evolution is incorrect, what accounts for all the bodies of animals discovered as fossils in the earth?

I will attempt to answer some of these questions.

Read on.

Chapter Twelve

THE ORIGINAL EARTH

he Ancient World

Let's assume that Genesis is correct. Let's accept the description of the world at the time of Adam and Eve. We understand that the earth was quite warm and without rain (at least in the Garden of Eden), being watered by a mist. This means that there was probably very little wind and rather high humidity. Warm and humid, like a greenhouse. There were neither summer nor winter. Everything grew well; crops came into harvest all year. It was a tropical paradise. Could the Earth have really existed in this ideal condition? If so, why is our world no longer like it once was? The Bible has some fascinating insights into these questions. Apparently, the Earth has radically changed since the time of creation.

We can make some assumptions about weather conditions based on what we know about the Earth, namely: that it assumed its present orbit 10,000 -12,000 years ago; that its axis was nearly vertical; that it was no more than 92 million miles from the sum. This geometry would yield the following conditions:

1. A uniform temperature distribution over the earth's surface;

2. Continuous winds of convection, moving from the equator toward the poles;

3. A lot of evaporation from the oceans, producing a dense layer of water vapor which would also move toward the poles;

4. Heavy precipitation near the poles, but none near the lower latitudes;

5. Subdued sunshine in the daytime, due to the vapor blanket; clearer skies during early morning; and an absence of rain.

Does this fit the picture from Genesis? Yes, it certainly does. It was a beautiful, tropical earth, brimming with verdant plant growth; a place where God chose to place his highest creative achievement: man. What a splendid creature he was! He was not only a physical masterpiece, but, as a spiritual being, was able to commune with God. He could reproduce, unlike the angels, and God designed him to manage his world, and thus, overcome Lucifer's rebellion.

This was the creation of Genesis. God declared that it was " very good," then he rested. Lucifer was certainly livid with rage!

The Mammals

We must not overlook the other fabulous creatures that God created. Among the last things created are the mammals (also created on Day Six) that is, having mammaries and able to breast-feed their young. These are a true marvel. Mammals are loving and they nurture their young; moreover, the babies are born live from the mother's body, and feed on milk. They are warm-blooded and relatively intelligent. God had a purpose for these creatures as well. Note his comments near the end of Genesis 1: 28.

And God blessed them and said unto them, Be fruitful and multiply, and replenish the earth, and subdue it. . .

What an interesting command! Replenish the earth (was it damaged?) and subdue it (was it untamed?). God intended man and the mammals to predominate, with man in charge! God designed a planet where He was honored, a place where precious life could flourish under the watchful eye of a capable supervisor. God truly loved his creation. All of it.

Mass Extinctions

A great number of animals have perished from the Earth. We are talking in terms of many millions of creatures, the majority of them large and quite robust, apparently, cut off in the prime of life. Remains have been found in huge mass graveyards, frozen solid. Some of these beasts were unmarked and lay right where they fell, their bodies well-preserved, often with fur intact and food still lodged between their teeth. Not just a few species have been found; the list is long. These were not gentle beasts, unable to resist an enemy. Among their ranks was the saber tooth tiger, larger than our African lion; as well as the fierce short-faced bear, an immense animal with long legs, capable of traveling at great speeds over long distances. These animals were big, fast and cunning and seemed to be thriving up to the moment of death.

So how did they die? Why did they die in such great numbers? This was a puzzle which Charles Darwin could not answer, even though he was aware of it. As archeologists made new finds, the mystery only deepened. For instance, in 1901 there was a gold-rush in Alaska. Miners, probing the beds of long-frozen rivers and streams, uncovered ghastly cemeteries of animal remains. They found hundreds of creatures, many piled together in great heaps, frozen solid, some with their mouths open and eyes still wide in fear. They were beasts which no man recognized. Scientists were brought to the scene to identify these extinct animals. They included the saber tooth tiger and short faced bear, noted above, along with woolly mammoths and mastodons, giant sloths, several types of camels, varieties of horses, a huge lion-like cat called Felix predensis, a gigantic type of buffalo, the Imperial Mammoth and the dire wolf. All of these were found together in a single location. After full examination, it was found that only two of this group are alive today, the common camel and the well-known horse. However, neither of these were found in North America when Columbus landed in 1492. Horses were brought by the Spaniards in the 1500's and camels arrived much later as circus attractions.

What an incredible and fascinating discovery! There were other such finds, including some that were made within the last year or two. A baby mammoth was found frozen solid, perfectly preserved in every detail, by a group of Eskimos in Siberia during 1990. The population of woolly mammoths in North America and Siberia was estimated to have been 10 million; yet not one is alive today!

Do you find this astonishing? I certainly do! All of these animals were more powerful than any we see today. They were rugged, well-adapted and very fit. According to Darwin, these were the fittest which should have survived, but they are gone. The list includes hundreds of different species and varieties. Does this sound like evolution to you? I think not.

Don't just take my word for it. Let's examine a couple of factual studies which deal with the disappearance of two of these animals, the Woolly Mammoth and the Dire Wolf.

The Mammoths

This was a member of the elephant family, only larger and covered with a heavy growth of hair. Its tusks were huge, and it had a massive trunk. The males were about 50% bigger than the African elephant, with larger legs and feet. They lived in herds of 1000 or more, much like the American bison. You can imagine that they were not easy to kill.

Mammoths were plant-eaters, grazing on grasses and leaves. We know this because of the ones found with grass still in their mouths. They died in groups of several hundred in one place. Their bodies were found on top of the ground, not as fossils. In 1799, European explorers found the first mammoth in perfectly preserved condition. Since then, thousands have been located and the story is always the same: The animals show few signs of decay or putrefaction. They died swiftly, probably within minutes, and their bodies froze rapidly. Russian scientists estimate there to be five million dead mammoths in Siberia.

In North America, mammoths were almost as numerous, living in herds across Michigan, Wisconsin and Minnesota. They apparently all died in a single day. These animals did not freeze, however, because these areas have warmer weather. The bones of several mammoths were found about 80 miles north of my home town of Port Huron, and are now on display at our Museum of Arts and History. These animals died at the same time as the mammoths in Siberia, as tested by radio-carbon dating. The year was about 3500 BC or 5500 years ago. This is the same age as the Niagara Gorge, as you will recall.

The Dire Wolf

Two species of wolves ranged across North America in the recent past, the familiar grey Timber Wolf, the long-legged creature seen in Walt Disney movies, and the Dire Wolf, a much bigger creature with enormous jaws. We have all seen a Timber Wolf in the local zoo. They are rather lean hunters, traveling in packs of ten or twelve. They weigh approximately one hundred pounds and are not aggressive unless hungry. This wolf is rather dog-like in nature; a bit lazy, quite intelligent, and basically shy. It is a predator, a silent hunter, but rarely dangerous to mankind.

The Dire Wolf was a different beast. It was big, weighing up to two hundred pounds with large feet and powerful haunches, making it a swift runner. It had larger teeth and stronger jaws than its cousin, the Timber Wolf. Dire Wolves traveled in packs of 50 or more, and were ferocious hunters. Further, there are signs that the Dire Wolf was clever and very aggressive. According to Dr. E. C. Pielou, Dire Wolves had

97

taken over two thirds of North America while the number of Timber Wolves was dwindling. Then, seemingly overnight, the Dire Wolf died off, leaving no survivors. The common Timber Wolf, less aggressive and clearly less fit, is still with us today. How could such a flip-flop of nature take place? Why and when would a flourishing species, bigger and stronger than the competition, suddenly become extinct?

The Dire Wolf remains have been tested and variously dated as being between 8000 and 5500 years old. That means that the Dire Wolf died about the time that the Mississippi Delta began to form: 3500 BC.

Let's face it, all of these countless extinctions and vast geological alterations point to a catastrophe of enormous proportions. Such destruction could only have been caused by a global disaster, one that swiftly eliminated many hardy species while carving huge new landmarks.

The Bible records such a catastrophe in the Book of Genesis. We commonly call it the Flood of Noah.

Read on.

Chapter 13

GOD'S MASTER PLAN

hat Makes us Human?

Man is far more than an intelligent animal. He is a creature with a living soul and a special identity. Man has a particular quality which sets him apart from other mammals: Spirituality. You cannot deny this, for every person has a built-in sense of worship. We all seek a higher purpose in our lives. Studies have demonstrated that animals' behaviors are centered around their present circumstances. Only man looks to the future, seeking some purpose for his existence. Even a brutal man like Adolf Hitler sought to link his actions with a higher purpose and calling.

Charles Darwin worked hard to deny the spiritual needs of people. He suggested that man's religious displays were comparable to animals' behavior patterns. Indeed, many creatures do show rituals in their behavior, especially when they are involved in mating. Darwin declared that religious pursuits were merely a sublimated ritual, which had evolved into a more complex form. He denied that man has a soul which is immortal.

We have been concentrating in this book on methodically attacking Darwin's theories. The steps of evolution did not take place, concluding that man did not evolve. Now, we are left with the following questions which remain unanswered:

What makes us human?

Why are we so different from animals?

What is the reason we are on the earth?

These are probing questions. How can we explain this gap between men and the other creatures?

The Vast Gulf

One of the most famous scientists of today is Richard Leakey, an anthropologist from Kenya, Africa. He discovered the famed fossil-remains of the Turkana Boy along the shores of Lake Turkana in 1984. These are fossilized bone-fragments taken from a bed of clay, dated as 1.5 million years old. It is important to note here that Dr. Leakey is a confirmed Darwinist. Yet he admits in his latest book, *Origins Reconsidered*, that there are enormous gaps in the fossil record. He even admits that Turkana Boy is not a missing link, nor have any valid missing links been found to date. Further, Dr. Leakey has made reference to the differences between humans and their closest relatives, the apes. Men are guided by a conscience; animals have no sense of right and wrong. For example, an ape will kill another animal even when it is not hungry. A male lion will kill another male, even its own son, in order to be dominant. Men do not normally behave like this. Dr. Leakey has been unable to explain this marked difference.

Thomas Huxley, who was one of Darwin's earliest disciples, also noted the obvious disparity between men and the other mammals. This variance worried him, since it seemed to contradict the tenants of evolution. If man had evolved, there should be greater similarity in behavior patterns. Huxley often referred to the 'vastness of the gulf between men and the brutes.' Recent studies of the apes and gorillas have confirmed their lack of conscience and/or higher thoughts; they live very much in the present, with no view toward personal improvement. Exactly what Huxley feared has been documented.

To this day, no scientist can explain why men are such abstract thinkers and organizers, while the other mammals lack this capacity.

This difference is God-given. Men were made to be superior, to plan and organize and to solve problems. Man was designed by a brilliant Creator, fashioned according to a

101

sacred and splendid blueprint. We were created to play a role which is meaningful and eternal. God placed within man an immortal soul which does not die. He gave each of us the ability to imagine great things, to have a grasp of complex matters, to make reasonable decisions. More than this, God also gave us the ability to talk directly with Him; we were designed to enjoy the Lord, not fear him. We were intended to be companions and trusted stewards for the Great God of the Universe.

Man was made in God's own image, and that is why we are different.

The Way It Was

We have already discussed the original sin and fall of Lucifer, who was one of God's earliest creatures. This angel, although powerful and privileged, wanted to be the Boss. He wanted to be 'like the Most High.' In doing so, Lucifer introduced sin and chaos into our Universe. At some point, he began to interfere with God's creative efforts, subverting some of the creatures the Lord had made (I referred to the serpent, you recall). As a result, Lucifer was ejected from heaven while God continued his efforts on the Earth.

To defend against further incursions by Lucifer, God made human beings. They were given perfect bodies and eternal life, in order to multiply and become the wise guardians of all creation. The earth was given to mankind, to supervise and improve, and, above all, to subdue. The last vestiges of Satan's work needed to be wiped out. This would be hard work, but the tools were at provided. Huge and powerful mammals were there, only needing to be organized and trained. What a grand and heroic assignment! Man was to bring the whole earth under God's control, in the role of resident managers. This is clearly spelled out for us in Psalms 8: 4-6.

4. What is man, that thou art mindful of him? And the sons of man, that thou visitest them?
5. For thou has made him a little lower than the angels, and crowned him with glory and honor.
6. Thou madest him to have dominion over the works of thy hands (the earth); thou hast put all things (creatures) under his feet.

There is the purpose for mankind, in a nutshell. Men were created as physical beings, not as swift or mobile as the angels, but able to take direct physical action. Earth was to be man's arena. We were given glory and honor, until Adam and Eve changed that. Still, man retained dominion over the earth, and command over the animals. We were admirably designed to be guardians, planners, builders, and supervisors; this was to be our service to the Lord. In return, man would live indefinitely, accumulating wisdom, acting as the sage captains for our planet.

Man has failed in this, his first and only assignment. We have slipped further away from the Lord with each passing generation. God could not allow men to live forever, since they had disobeyed. Sin and death entered the picture. Men would die physically, but what about his soul, which would go on for all eternity? Would the spiritual side of mankind be forced to dwell with Lucifer and the fallen angels?

God at Work: Salvation

We should mention the rescue of creation by Noah. He and his family became the ancestors of all humans. These were the Cro-Magnon people who survived, not because they were the fittest, but because God sent a way of escape from the coming disaster. The method he chose was a boat, called the Ark, which was designed by God and built by Noah. The craft was over-sized for its purpose, able to carry several hundred people and thousands of animals. Why had so much extra room been provided? Perhaps God hoped that other men would repent and join Noah on board, but none came. Still, the Lord waited patiently so everyone would have a chance to escape destruction. This is noted briefly in I Peter Chapter 3: 20.

. . . when once the long-suffering of God waited while the ark was a-building, wherein only a few, that is, eight souls were saved by water.

The ark took many years to build, during which God suspended his judgment. Yet only a relatively few were saved from death, eight souls in all. God saved everyone of those who came into the ark, including the animals. The world went on eating and drinking, thinking Noah was crazy, as Jesus noted in Matthew 24. Then the Flood came and swept them all away.

Eight people, descendants of the original creation, rode to safety on the waves of a great flood. Pairs of selected animals also were saved, spared from death by the obedience and labor of a single man and his family.

This was salvation, as man was spared by a God unwilling to kill the righteous in order to destroy the wicked.

God provided a way of escape, which He himself designed (the Ark) more than 5500 years ago; and by the hands of Noah and his sons, his creation was spared. By the intervention of one righteous man, salvation was achieved.

Salvation for Today

In the same manner, the Lord provided a way to salvage the souls of man by means of a Savior. He, himself, would act as Noah did, intervening on behalf of the all of creation. He

105

would succeed, in spite of towering odds against him. This man would provide a means of restoring men to spiritual wholeness, according to the plan He had laid down in distant ages past. In other words, God intended to restore men to that elevated place which they had held in beginning, as we can read in the summary of Ephesians Chapter 1:9-10.

9. *(God) having made known unto us the mystery of his will, according to his good pleasure, which he hath purposed (decided) in himself.*

10. *That in the fullness of times, he might gather together in one, all things in Christ; both those things in heaven and on earth, even in him.*

So God has made known to us his original plan, which was a mystery for many years. He decided in his own mind long ago to bring all things together under One Ruler, the man Jesus Christ. This would include all things, both in heaven and on earth, visible and invisible.

Christ is that one head, the Royal Prince and Ruler of the Universe. Eventually, every knee will bow before him. All the angels will bow, both good and evil. Satan himself will bow. Every man of every age will bow to him. Even Charles Darwin, the High Priest of the Cult of Evolution, will drop to his knees, and call him Lord.

We who are alive are more fortunate than Darwin, for his privilege of choice is gone forever. He must bear the consequences of the theory which carries his name; Darwinism will not bring his salvation. His fate is sealed.

On the other hand, you and I are still living. We can serve God as He intended; we are still able to choose to follow Jesus Christ, the Lord and Master of the Universe.

Each person must make this choice; perhaps you have already done so.

I can speak only for myself; happily, I have chosen Christ, who promises an eternal existence with the God of the Universe. This offer is available for you, too.

As I close this book, I must utter a sad parting word to the originator of evolution.

Goodbye, Darwin!

APPENDIX - BIBLIOGRAPHY

A. BOOKS

1. *Einstein's Universe* by N. Calder, Penguin Books, 1987.

2. *Relativity: The Special and General Theory* by Albert Einstein, Crowne Publishing, 1989.

3. *Worlds in Collision* by I. Velikovsky, Simon and Schuster, 1978.

4. *Earth in Upheaval* by I. Velikovsky, Simon and Schuster, 1977.

5. *Genesis and the Big Bang* by G. Schroeder, Bantam books, 1990.

6. *Biblical Creationism* by H. Norris, Baker Books, 1993.

7. *How the Leopard Changed it's Sports* by B. Goodwin, Charles Scribners and Son, 1994.

8. *Origins Reconsidered* by R. Leakey, Doubleday, 1992.

9. *God and Science* by C. Henderson, John Knox Press, 1986.

10. *After the Ice Age* by E. C. Pielou, University of Chicago Press, 1991.

11. *The Origin of Species* by C. Darwin, New American Library, 1958, 1993.

B. PERIODICALS

1. *Greenland Ice Shows Climate Flip-flop Science News*, September 26,. 1992.

2. *Is Mars a Giant Asteroid?*, New Scientist, July 13, 1991.

3. *Darwin May Mounder on the Great Barrier Reef*, New Scientist, October 20, 1990.

4. *Edwin Hubble and the Expanding Universe*, Scientific American,

July 1993.

5. Flap Over Earth's Magnetic Flips, Science News, June 12, 1993.

6. Ice-cores Shows Speedy Climate Change, Science News, June 5, 1993.

7. Modern Humans Linked to Single Origin, Science News, April 2, 1994.

8. Satellite Uncovers Ancient Arabian River, New Scientist, April 3, 1993.

9. Melting Ice Submerged Migration Route to Alaska, New Scientist, 15 May 1989.

10. Rocky Relies: The Near-Earth Asteroids, Science News, February 5, 1994.

11. Impact Wars: The Debate Over Killer Comets, Science News, March 5, 1994.

12. Earth's Orbit and the Ice Ages, Scientific American, October 1988.

13. End of the Ice Ages?, New Scientist, June 17, 1989.

14. Frozen Baby Mammoth, New Scientist, December 10, 1988.

15. When Rhinos Roamed the Alps, New Scientist, October 26, 1991.

16. Broken Teeth: Tale of Sabertooth Tiger, New Scientist, August 1993.

17. Dwarf Mammoths Outlived the Ice Ages, Science News, March 27, 1993.

18. Apes and Humans: Curious Kinship, National Geographic, March 1992.

19. The Neanderthal Mystery, Time Magazine, March 14, 1994.

20. Dinosaurs: They Roamed the Earth, National Geographic, January 1993.

21. *The March Toward Extinction*, National Geographic, June 1989.

22. *Mega-Bergs Left Scars on Atlantic*, Science News, August 20, 1994.

23. *Staggering Through the Ice Ages*, Science News, July 30, 199

Ordering Information

Would you like to order additional copies of this book? Key Publishing Company will ship copies direct to your door. All books are shipped by U.S. mail unless otherwidse specified.

Price Schedule

1 — 3 copies	$7.95 each
3 — 6 copies	$7.50 each
More than 6	Suitable discounts available for lot purchases

Mailing/Handling

Please add $2.50 for first copy; $1.75 additional for second through sixth copy, to cover packaging mailing costs. U.S. funds only — do not send foreign currencies.

Payment

We accept personal checks, money orders, or Visa. If you are using a special purchase order, make it out for the full amount of sale. Personal checks accepted in U.S. funds only.

Key Publishing Company

3240 Pine Grove Avenue
Port Huron, MI 48060
Phone: 810-984-5579
Fax: 810-984-5595